D1090441

May Gibbs
T.M.
Counting
Book

CORNSTALK
PUBLISHING

CORNSTALK PUBLISHING
Cornstalk, an imprint of HarperCollins*Publishers*

25 Ryde Road, Pymble, Sydney, NSW 2073, Australia
31 View Road, Glenfield, Auckland 10, New Zealand

Angus & Robertson edition first published in Australia in 1993
This Cornstalk edition 1995

ISBN 0 207 18829 7

Printed in Hong Kong

6 5 4 3 2 1
98 97 96 95

1 one

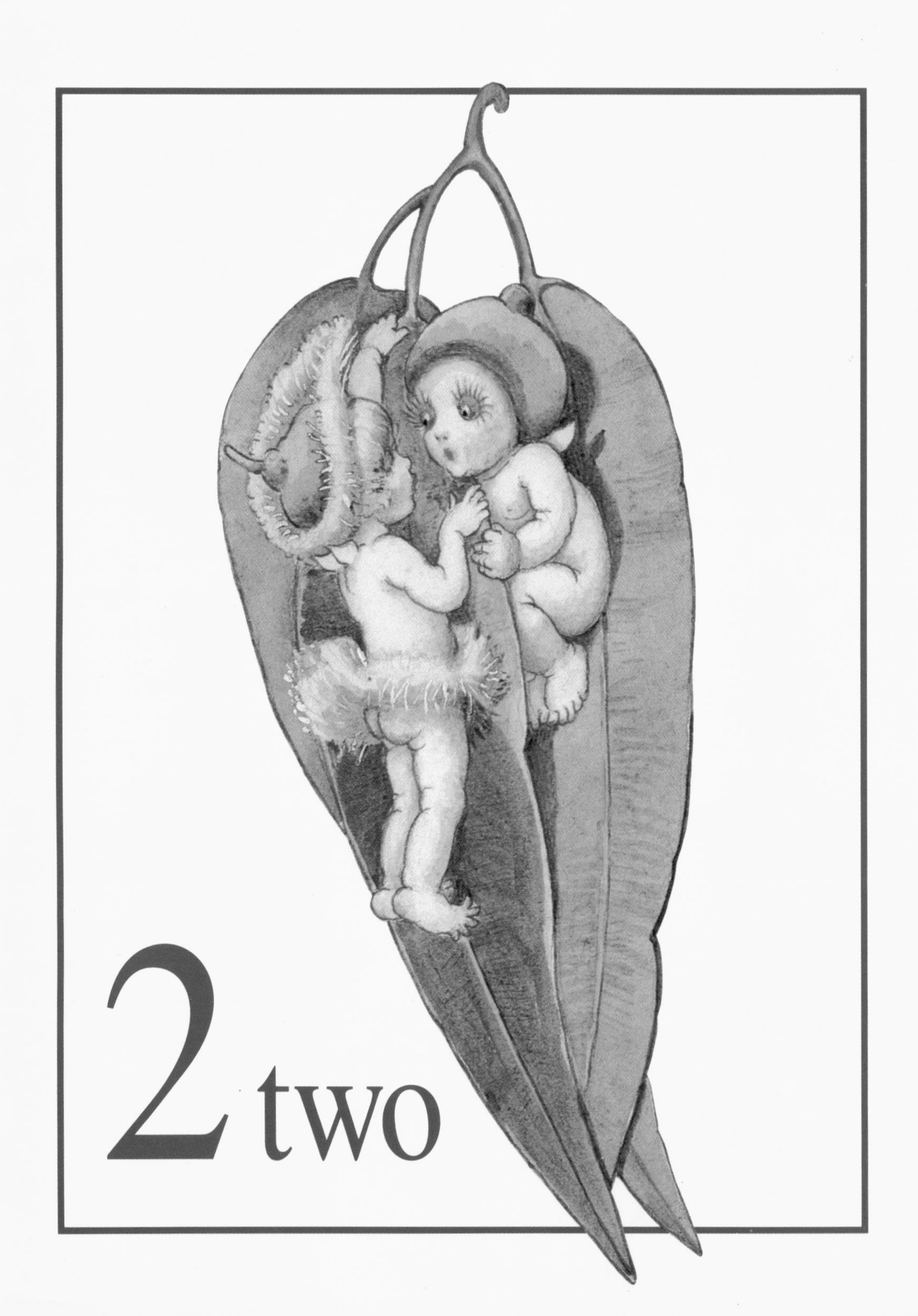
2 two

3 three

4 four

5 five

6 six

7 seven

8 eight

9 nine

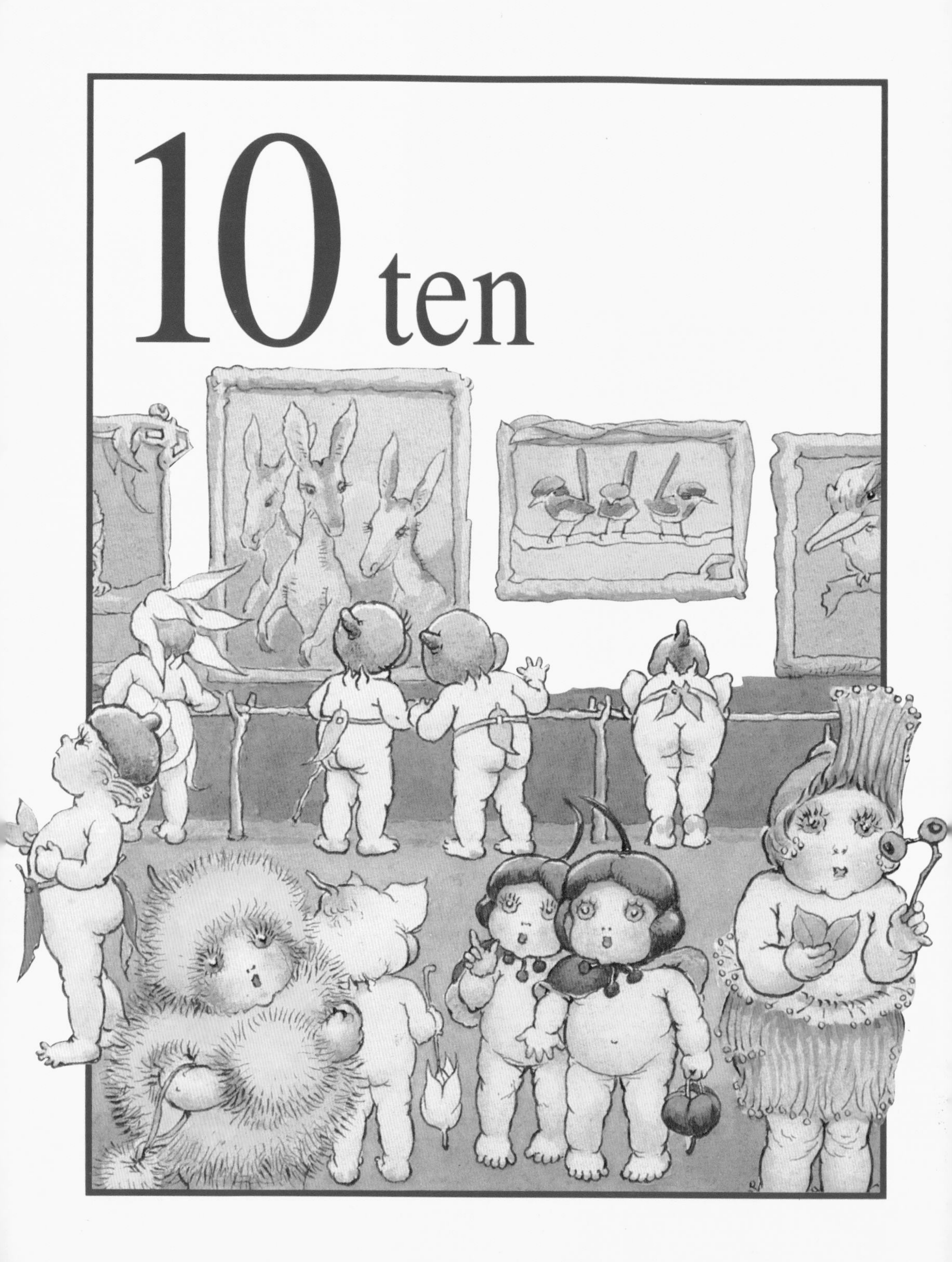
10 ten

11 eleven

12 twelve

13 thirteen

14 fourteen

15
fifteen

20 twenty

30 thirty

40 forty

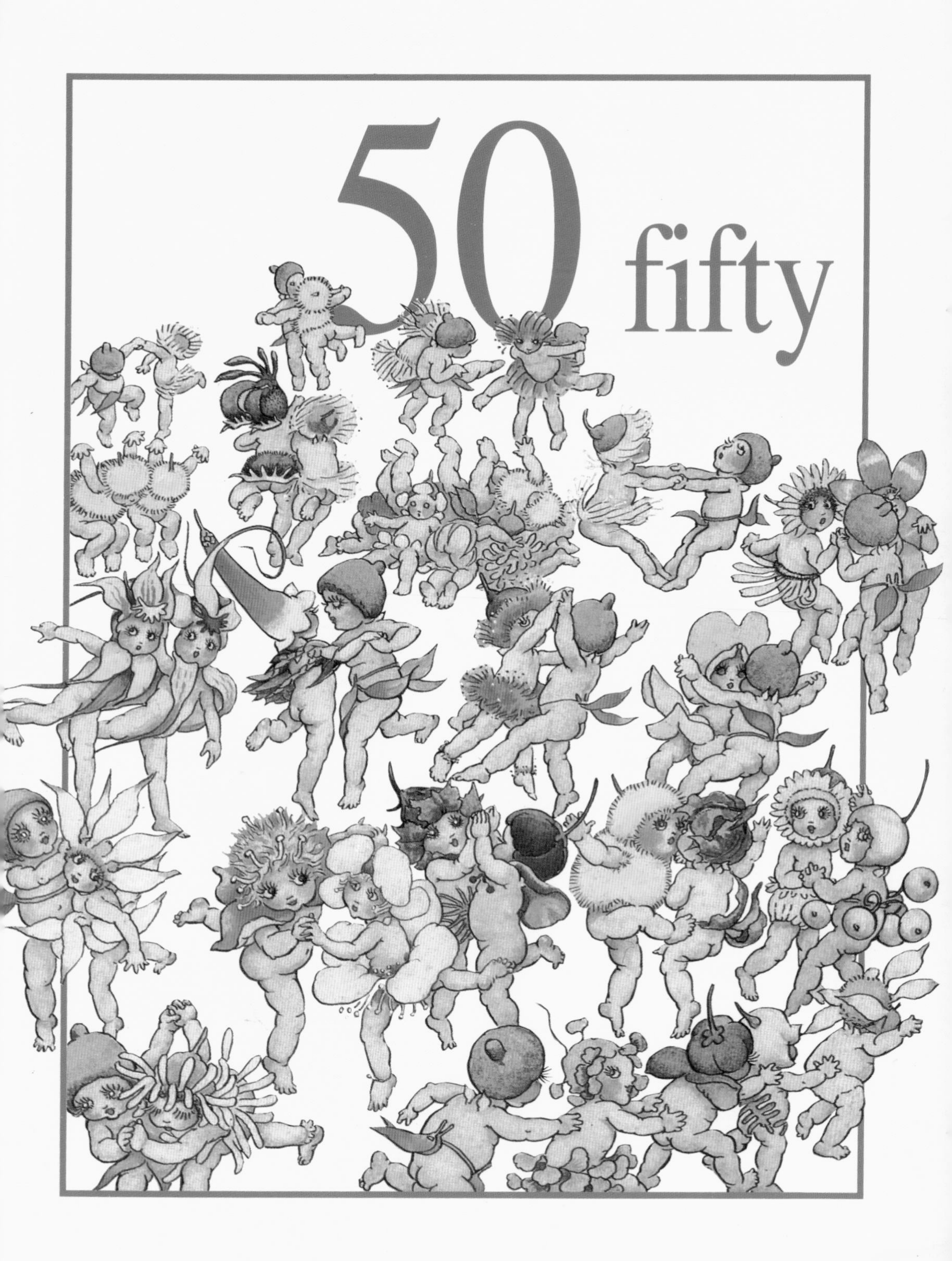
50 fifty

100
one hundred

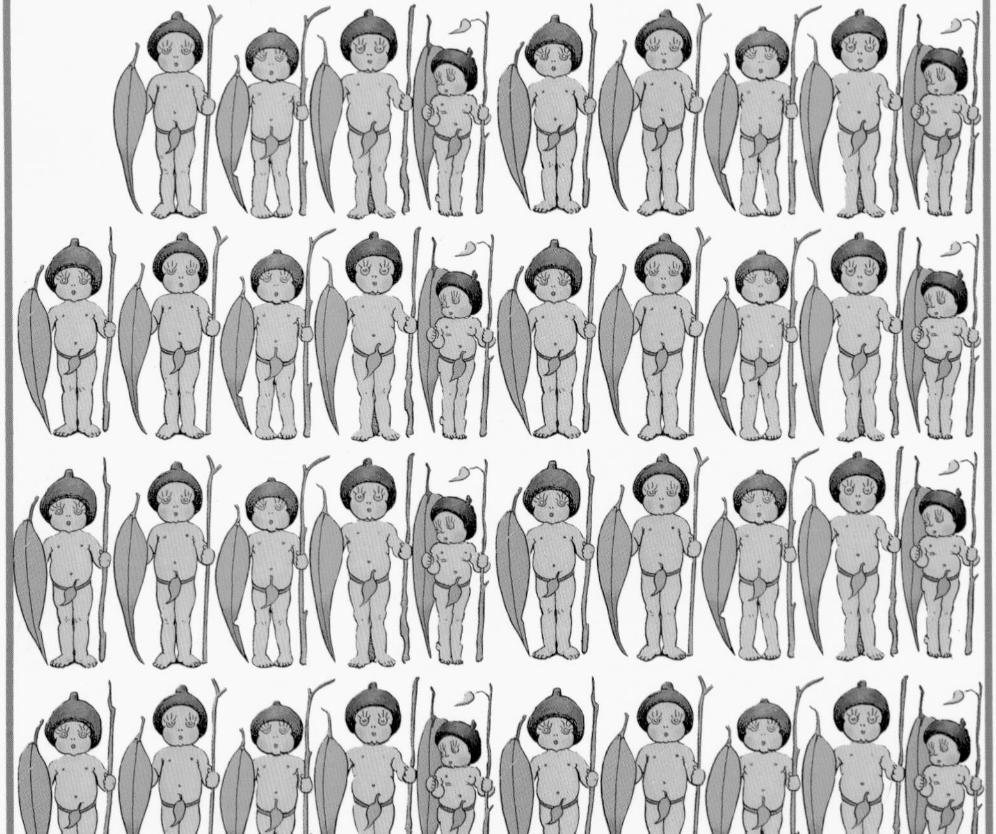

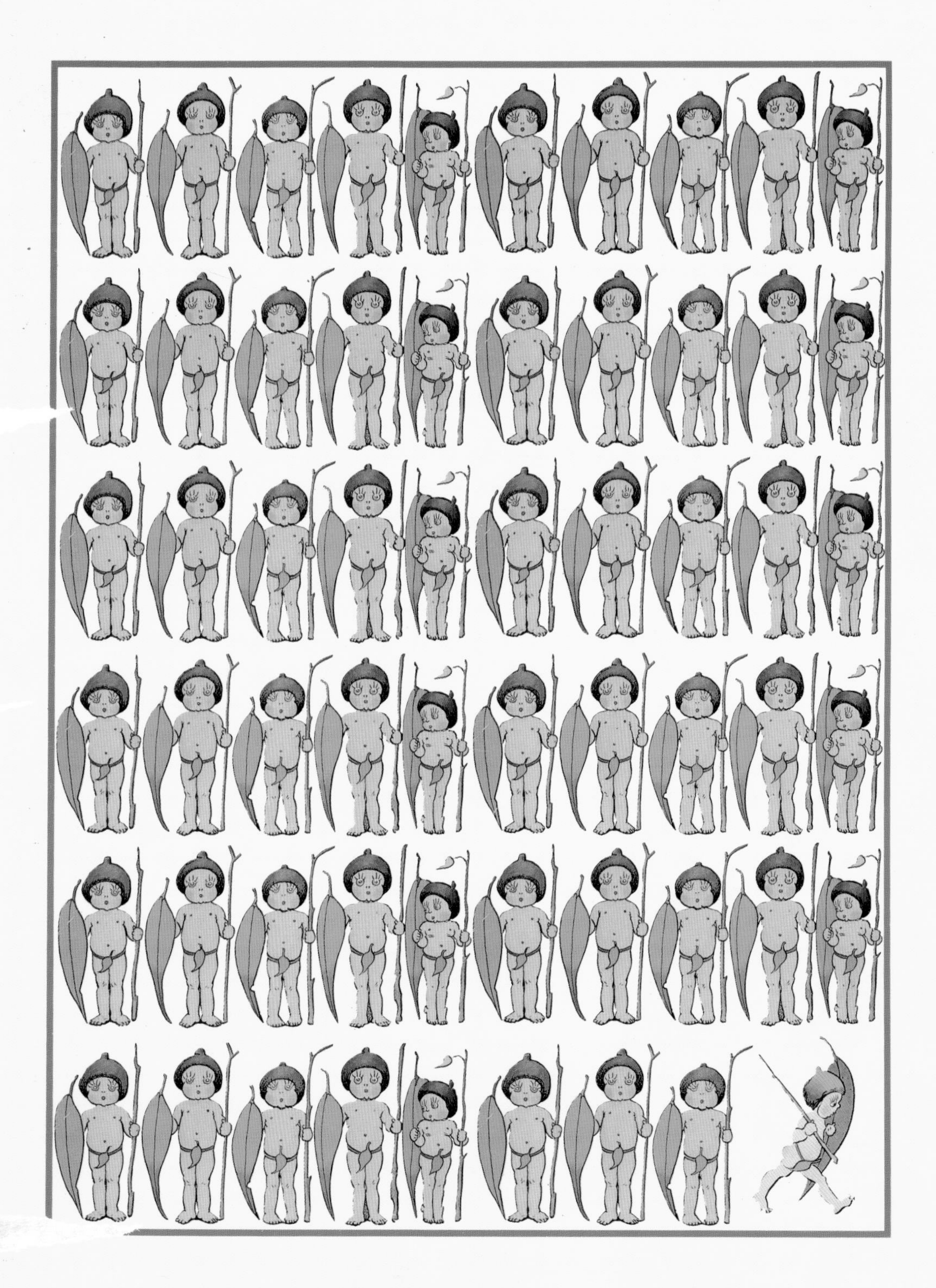